New Zealand Bird Views

SWANS AND GEESE

Raewyn Adams

Published 2013
Te Puke, New Zealand.

ISBN 978-0-473-26893-0
Copies of this book can be purchased from www.lulu.com or your favourite online bookseller
Author contact radams.photo@gmail.com

Front cover: Black swan at McLaren Falls Park, near Tauranga. May.

CONTENTS

SERIES INTRODUCTION

The concept of this series is to present and appreciate the birds that people are likely to view around them. These are the common birds that are around us every day but we fail to really see because they are always just "there".

The intent of this series is to tell the bird's story in photos. In this format the reader can see more of the bird's life and gain an understanding and appreciation of its place in nature and how it interacts with the world it lives in – its relationships with other birds, the environment and also with people.

When you get to know them, birds are fascinating animals. They have complex social structures, sophisticated communication systems and are very hard workers. They are also very beautiful. The feathers of even the plainest birds have amazing textures and intricate patterns.

I have provided some basic factual information. For those wanting more depth there is a starting point in the list of further information resources at the end of the book. The rest of the book consists of my personal observations, illustrated with photographs as much as possible.

I hope you enjoy these views of the birds around us.

SWANS AND GEESE IN NEW ZEALAND

Three species of swans and geese are commonly seen in New Zealand:

- **Black swan** (*Cygnus atratus*). Approximately 100 birds were introduced to New Zealand from Australia in 1864-1868 as game birds and it is likely that some have also self-introduced. Now found throughout New Zealand, it is protected except for the game bird season. Approximately 50,000 birds. 110-140cms long.

- **Canada goose** (*Branta canadensis maxima*). Introduced to New Zealand in 1905 from northern and central USA. Found extensively throughout the South Island and in increasing numbers in various North Island locations. Previously a game bird, since 2011 it has been completely unprotected. Approximately 60,000 birds. 85-95cms long.

- **Domestic/Feral geese** Most domestic geese in New Zealand are breeds that were developed from the Greylag goose (*Anser anser*) over long periods of time, sometimes thousands of years. Found throughout New Zealand, especially in urban lakes and near farm pasture and ponds. Not protected. Population unknown but there are likely to be a few thousand. 80cm long.

Other species that may be seen occasionally are listed at the end of this book.

Above: Two feral geese at left, Canada goose in the foreground, immature black swan behind. Below: The same four birds. Matata, Bay of Plenty. January.

BLACK SWAN

The black swan was introduced as a game bird in the 1860s. It is also believed that birds have self introduced both before and after this, which means it could be considered native. Its legal status is as a game bird and it is protected except when hunted by licensed hunters during the game bird hunting season. Control of numbers is the responsibility of the Fish and Game New Zealand Council.

A typical habitat for black swans - an urban lake. The swans cohabit with ducks, pukekos, geese, etc., and line up with the other birds when food is being provided. Both photos Gordon Carmichael Reserve, Tauranga. January above, June below.

Black swans are found on lakes and large ponds throughout New Zealand, sometimes in large numbers and sometimes just a pair or two. The freshwater environment is where the swans breed from late winter through spring. Non-breeding birds can also be seen on some estuaries where they congregate in large flocks. Estuary numbers are greatest when the birds are moulting and flightless, but some birds remain on the estuaries all year round.

Swans at urban lakes and ponds are often well used to people and will come to be fed along with the ducks and other semi-tame waterfowl. Male swans guarding their young may be aggressive towards other birds and people, but apart from that are tolerant of the presence of other species.

Males and females look alike, but the male is larger. He is called a cob, from the Middle English word cobbe, meaning a leader. The female is called a pen, possibly related to the Middle English word penne, meaning quill. The young babies are cygnets. They mature to a dark brown version of their parents, gradually darkening to black before breeding at 2-4 years old. The life span is about 15 years.

A group of black swans on estuary mudflats. Tauranga Harbour. February.

Black swans on the estuary are nervous and will fly away as soon as they realise a person is approaching. Tauranga Harbour. March.

Swans feeding and relaxing at high tide. Te Pua Point, Far North, April.

Two classic views of black swans. Above: McLaren Falls Park, near Tauranga. May. Right: Lake Rotoiti, near Rotorua. September.

Swans on land look ungainly and out of balance. Lake Taupo. October.

A close up of the head shows the intensity of the colour of the red bill and the contrast with the white tip. McLaren Falls Park, near Tauranga. May.

Elegant on the water. McLaren Falls Park, near Tauranga. May.

Relaxed with one foot held above the water. McLaren Falls Park, near Tauranga. May.

The full extent of the white feathers is usually only displayed when the birds beats its wings. Hamurana, near Rotorua. October.

View of the underside of the wing. McLaren Falls Park, near Tauranga. April.

Take-off is not easy for such large birds - they patter noisily across the water for about 25 metres, or even up to 50 metres, before becoming airborne. Flight is not possible if there is insufficient clear water. Lake Rotorua. March.

Immature swans have less intense colour in their black feathers, plus black tips on their white feathers. Gordon Carmichael Reserve, Tauranga. March.

CANADA GOOSE

The preferred habitat of the Canada geese in New Zealand is pasture next to a lake but the birds may turn up at any waterway of reasonable size, including both back country and urban lakes, large ponds, coastal lagoons and estuaries.

About 65% of the New Zealand population is in the South Island, mainly east of the alps with increasing numbers in Fiordland and Southland. They are also becoming more common in central North Island locations from the Wairarapa to the Waikato and Bay of Plenty areas.

Canada geese are usually seen in flocks, which can be large. Truly wild birds are shy and will avoid contact with people, but those in an urban lake environment may become reasonably tame. Some line up with the other waterfowl for offerings of bread, but not quite as readily. Over time, they may learn to recognise the individual people who feed them regularly and can be quite selective about whom they approach.

Males and females are alike but the gander (male) is larger than the goose (female). The newly hatched young are goslings. The life span is about 24 years.

A group of geese together are most often regarded as a flock, but correct collective terms for geese are a gaggle for a group of geese on the ground, and a skein when they are in flight.

The colour is uniform – always a light brown body, black neck and white chin marking. All other geese look different so identification is always clear.

Above. At an urban lake, a pair of Canada geese and their goslings are lined up for bread with ducks and a pukeko. Only a few geese at this location come to be fed but that number will probably increase as the goslings learn to be comfortable with people and continue the cycle into future generations. Below. Resting on the lake. Gordon Carmichael Reserve, Tauranga. December above, June below.

Swimming past Motuopuhi (Rat) Island at high tide. Tauranga Harbour. June.

Sharing low tide with South Island pied oystercatchers. Waimapu Estuary, Tauranga Harbour. June.

Canada geese grazing a farm paddock adjacent to a lake. Lake Okareka, Rotorua. September.

When on land, the Canada goose has a much more upright stance than the black swan. Gordon Carmichael Reserve, Tauranga. August.

The white cheek pattern is uniform in all Canada geese. Gordon Carmichael Reserve, Tauranga. August.

Outstretched wings show us the texture of the feathers in a way we don't otherwise see. Gordon Carmichael Reserve, Tauranga. July above, March right.

Two views of Canada geese in flight. The front view at right shows how the white chin strap circles the face. Both Lake Okareka, Rotorua. November.

Flying overhead in a classic "V" formation. Waimapu Estuary, Tauranga Harbour. June.

"Landing gear down" approaching the pond. Gordon Carmichael Reserve, Tauranga. February.

Canada geese showing their grace and beauty on the water. Both Gordon Carmichael Reserve, Tauranga. December above, August right.

DOMESTIC/FERAL GOOSE

Geese originating from the European greylag goose have been domesticated for at least 2,000 years. There is a large variety of breeds both in New Zealand and overseas. These are bred by poultry fanciers, kept by farmers and lifestylers and some escape or are released to the wild to form feral colonies.

Domestic geese are bred for their eggs, meat, fat and/or down. They can also be kept as pets, used to control weeds and make good "watch" animals.

A few were brought here by Captain Cook and later acclimatisation societies imported birds but these did not survive. The population of feral geese that we have today has come from domestic birds imported and bred privately. It is not always clear whether birds are wild or owned. Those found on a public lake are likely to be feral. Those found on farmland roadsides may belong to the landowner or neighbours.

The breeds that may be seen in New Zealand include:

- **Buff Back.** A medium sized pied goose with buff head, back and thighs on a white body. This colour pattern is also known as saddleback. Both sexes coloured. Blue eyes. Male up to 7.7kg.
- **Embden**. A large pure white goose (both sexes) Light blue eyes. Male up to 15.4kg.
- **Pilgrim.** A light-weight goose. Males are white with blue-grey eyes and females are grey with brown eyes. This sex-linked colour or auto-sexing is also found in the larger pied **West of England** breed and its smaller cousin the **Shetland**. The male Pilgrim weighs up to 6.35kg.
- **Roman**. A small, lightweight, pure white goose with light blue eyes. Male up to 6.3kg.
- **Sebastopol.** Another completely white goose, but with unmistakable long, soft, curly feathers. The wing feathers lack a shaft which means that this bird is flightless. It is therefore unlikely to be seen in the wild. Bright blue eyes. Male up to 6.3kg.
- **Toulouse.** A very large, heavy, grey goose (both sexes) with brown eyes. The main distinguishing feature is a prominent dewlap under its chin. Male up to 13.5kg.

Internationally there are many more breeds of geese, some of which may also be present in New Zealand.

Feral and farm grown geese may be crosses of any of these breeds so the birds seen may show a variety of colour patterns. Some may still show the traits of their pedigree ancestry. Others are just a random mix.

Many feral birds seem to have evolved from the lighter breeds, possibly because they are better fliers and escape captivity more easily. For example, although the Pilgrim is a fairly rare breed internationally, many feral and farmyard birds in New Zealand show characteristics of Pilgrim ancestry, including the sex-linked gene. This gives rise to the common misconception that all male feral geese are white. Most are, but it's not necessarily so.

Even when seen in feral populations, these geese are essentially domestic birds. After a few generations of living wild they may become suspicious of people, but they usually still frequent areas where people are not far away. They prefer to graze on pasture and are less likely to be seen in truly remote areas.

Feral geese. Their facial markings are as for Pilgrim but they are pied, not all grey. Hamurana, near Rotorua. August.

Two grey domestic geese on a farm. Adjacent to McLaren Falls Park, near Tauranga. April.

Pied feral goose showing the underside of its wings. Hamurana, near Rotorua. October.

Domestic geese usually seem at home on the ground, but many can, and do, fly. Whakatane. July.

Domestic goose showing the characteristic curly feathers of the Sebastopol breed. The slight colour in some of the feathers is acceptable in a young bird. Katikati. October.

A lone white feral goose - a male Pilgrim or a white goose of another breed? Kaikoura. March.

Two pairs of feral geese. The females are pied. The males are white. Lake Pupuke, Auckland. April.

The rest of the flock that the geese above right belong to. The pied females have a variety of markings. Lake Pupuke, Auckland. April.

Two pied feral birds. Hamurana, near Rotorua. Above September, right July.

Grey feral goose. Lake Rotorua. March.

A Canada goose swimming between two feral geese on an urban lake. Tinwald. March.

FOOD

Geese and black swans are primarily herbivores although there are differences in their feeding habits. Geese mainly graze terrestrial pasture while swans mainly feed on submerged plants. The habitats can overlap – swans can sometimes be seen eating grass, and geese will feed while swimming.

When living on lakes, geese and swans eat plants such as *Egeria, Elodea* or *Potamogeton.* The long necks of the swans allow them to access plants up to about a metre under the water, and they can often be seen completely up-ended as they strive to reach plants below the surface. Insects and other small animal life will be eaten along with the plants pulled out of the bottom mud. Canada geese also feed in this way, although generally they prefer pasture grass.

Domestic and feral geese occasionally eat while swimming but overall spend much more time feeding on grasses, seeds and other terrestrial plants plus a few of the snails and insects that may accompany them.

Grazing geese eat significant amounts of grass and a large flock can impact on farm feed management. Three Canada geese eat about as much as one sheep.

Geese have a very inefficient digestive system. With no teeth to chew the food, it is ground up in the gizzard and the juices extracted there. It then passes onwards through the intestinal system and is finally expelled as soft pellets like semi-chewed grass.

When feeding on estuaries black swans and Canada geese graze on the *Zostera* seagrass meadows that are exposed at low tide.

Canada geese grazing on a farm paddock. Lake Okareka, Rotorua. September.

While grazing the lake bank, this swan frequently paused for a mouthful of water. Grass is much drier than the submerged plants it normally eats. McLaren Falls Park, near Tauranga. May.

Canada goose reaching for plants well under the surface while a coot rests nearby. Lake Rotoiti, near Rotorua. March.

A more familiar sight – the tail end of a black swan reaching for the bottom. Lake Rotoiti, near Rotorua. October.

Canada geese feeding on a harbour. Waimapu Estuary, Tauranga Harbour. June.

When grazing the feral goose bites grass using the serrations along the edge of its bill. Lake Rotorua. April.

Feral goose feeding while swimming. The facial markings of this goose look like the the sex-linked Pilgrim breed but the blue eyes and saddleback colour pattern are more consistent with a breed such as the Buff Back. Hamurana, near Rotorua. July.

The seagrass that the birds eat on the estuary. When feeding in the salty environment of the mudflats swans also need to seek out supplies of fresh water. Waimapu Estuary, Tauranga Harbour. May.

Black swan feeding Egeria densa *to its family.* Lake Rotoiti, near Rotorua. June.

SOCIAL LIFE

For these birds, "social" life really means social. Although they can be found in individual pairs at times, they are most often seen in groups. Black swans and Canada geese form large flocks that number in the hundreds or even thousands of birds. Family groups of swans include the previous year's young plus non-breeding young adults up to about four years old and a few older birds. Within the large flocks the association of pairs and family groups is usually long term.

Canada geese communicate by honking, including while in flight.
Lake Okareka, Rotorua. September.

Black swans and Canada geese are very mobile, readily flying from place to place, especially when young. Swans are large heavy birds that need a lot of energy and clear water to get airborne, but once in the air they are strong fliers.

The social aspect of their lives together is also demonstrated with loud vocal communication. Domestic geese make good "watch" animals and both swans and wild geese tend to be very noisy.

They honk or hiss with aggression, call softly to each other when feeding or relaxing and call loudly to signal their intentions to fly. Within a group of Canada geese, one bird may be heard to call. Any response will be followed by another call, and another. In a few minutes the noise will be quite raucous then suddenly all the birds take off at once.

Group of feral geese walking together. Lake Rotorua. February.

Non-vocal communication is also important. Black swans either show or hide their white feathers to communicate different messages. Immature swans have fewer white feathers and with less contrast between the white and dark grey their displays are more muted than those of the adults.

A meeting of all three species can be amicable or can become two too many. In this case, two feral geese appeared to provoke the young swan. Later, one of the Canada geese chased the feral geese back to the water. Above and below Matata, Bay of Plenty. January.

When breeding, male swans will defend their territory or their mate very fiercely. Displays at this time involve a lot of white feathers being shown. In contrast, during courtship the white feathers are tucked away out of sight.

Acts of aggression can involve chasing the intruder away, or can be more complex rituals designed to intimidate the rival. The subordinate bird usually backs off, but a real fight can begin if it doesn't. The winner then uses a display of 2-3 wing beats that maximise the impact of the white feathers and intimidate other birds in the area to back off further.

Pairs may greet each other vocally and also perform a greeting display, especially following rival birds being chased away.

Geese also have vocal greetings and gestures. They have strong permanent pair bonds, including displaying what appears to be grief if one dies. In the wild they live in pairs or in family groups. Domestic geese learn to greet "their" people, even when they have been away for a few weeks.

They lack extreme courtship rituals because the pair bond is permanent and doesn't require them.

In the breeding season the male is possessive over his mate and will chase other geese away. If they end up in a fight, each grabs the other's feathers at the base of the neck then they beat each other with their wing spurs. The winner is the one that doesn't give up and shows his triumph with a raised wing display near his mate.

Outside the breeding season Canda geese live together harmoniously in large groups at various locations such as lakes or estuaries. At Gordon Carmichael Resserve, Tauranga, 50-100 birds regularly come and go all winter (above left). At Waimapu Estuary, Tauranga Harbour, there were three groups of about 100 birds each (above right). Both photos June.

The peaceful cohabitation changes when spring approaches. The two birds at left flew in and landed near a group that were grazing. The two birds at right stopped grazing and greeted the newcomers with what appeared to be aggression. Both photos Lake Okareka, Rotorua. September.

Right through summer and into autumn Canada geese can be quite territorial. Spats that start out with gestures (above left) can end up being a full on chase (above right). Both photos Gordon Carmichael Reseve, Tauranga. March.

Black swans have a similar behaviour pattern. The calm standover (above left) became a full on fight within a few seconds (above right). Both photos McLaren Falls Park, near Tauranga, Tauranga. April.

Alert and ready for a fight - head held high with bill vertical, white feathers displayed. McLaren Falls Park, near Tauranga. April.

The winner signals his victory with this raised head gesture accompanied by several controlled wing flaps that show the white feathers to best advantage. McLaren Falls Park, near Tauranga. April.

Aggression can also be aimed at other species, sometimes for no apparent reason. In this case the black swan chased a paradise shelduck. Lake Okareka, Rotorua. November.

This Canada goose suddenly took exception to the presence of the pukeko and chased it about 15 metres to the shore. Gordon Carmichael Reserve, Tauranga. June.

The pair bond is shown when birds greet one another with this greeting display. It is slightly similar to the male's victory display but is less extreme. Lake Rotorua. April.

Geese and swans can often be seen swimming in pairs. Gordon Carmichael Reserve, Tauranga. December.

When not actively breeding, all three species swim in groups with others of the same species. Here, a group of feral geese are swimming together. McLaren Falls Park, near Tauranga. May.

The photos on these two pages are a courtship sequence from a pair of black swans. Many aspects of the ritual are similar to other waterfowl with the main difference being the size of these birds.

Photos 1-3. For several minutes the birds alternately dipped and raised their heads, slowly and more or less simultaneously. Their heads were close together and their necks appeared intertwined.

Photos 4-5. The female positioned herself underneath the male with her neck stretched out and he took hold of the feathers on the back of her neck in typical waterfowl fashion.

Notice that there is no display of white feathers – they usually communicate aggression.

Photo 6-. The female rose up from the water in a way that appeared to tip the male off backwards.

Photo 7-8. A display similar to a greeting gesture was followed by the birds swimming in a circle with their bills pointed down.

Photos 9-10. Finally both birds bathed and male completed the ritual with wing-beating which effectively displayed his white feathers to any other males in the vicinity.

Both pages, Lake Okareka, Rotorua. September.

NESTING

Canada geese, domestic geese and black swans are all very family focussed. They maintain long term to permanent pair bonds and actively look after their young.

Immature black swans tend to live in large flocks in places such as harbour estuaries. When they reach 2-4 years old they may settle in a suitable fresh water breeding location and start to pair up and breed. Only some of the birds at a site breed each year.

Nests may be seen at almost any time of year, but the main season is through late winter and spring. Birds nesting individually start as early as July and may also raise a second brood. Birds living in colonies at lakes start breeding when summer water levels start to drop, possibly from September or later.

The nest is a platform built above ground on an island in a lake, or may be a floating raft on the water. 3-8 pale green eggs are laid, one each day. Both parents share the incubation which lasts 40-42 days.

In large colonies, nests may be very close together – sometimes only a metre apart. Dominant pairs take the best nest sites and may harass subordinate birds which in turn form a loose colony that provides more protection from this harassment.

When hatched, the cygnets leave the nest within two days. Sometimes one parent cares for the first hatched young while the other continues incubating the rest. When all are hatched, the family comes together or may join in with a larger group of young from several pairs to form a mixed family group.

Geese have some similarities in their nesting pattern but there are also differences; the main one being that only the female incubates the eggs. The male's role is to protect the area around the nest. After the eggs hatch, he stays with the family and helps to care for the young.

Canada geese usually lay 2-10 eggs in a simple hollow on the ground during September and October. The eggs are laid at a rate of one per day then incubated for 27 days. A few birds will have a second brood, possibly as late as January.

The female selects the nest site on bare ground or pasture, but near vegetation that will help to hide it. It may be a single nest, or one within a loose colony. The selected site needs to provide a clear view of the surroundings to enable approaching predators to be seen.

The nest is lined with down and when left unattended it will be covered in order to hide it.

The newly hatched goslings leave the nest within a day. They are brooded by the female and guarded by both parents. Or one pair of birds may care for a whole group of young from the wider colony.

Domestic geese also nest on the ground, preferably near water and under a tree or shrub for cover. The nest itself will be a simple hollow, lined with down.

Up to 9 (and sometimes more) eggs are laid at 34 hour intervals There is just one brood per year, usually from August to September but possibly through to December. Incubation is 28 days.

Feral goose incubating her eggs. This nest was about 2 metres from the water and partially protected by a low shrub.

A close up shows the down intermixed with the grasses and sticks to make a soft but fairly robust nest.

The only defence the sitting bird has against intruders is to hiss and try to chase them away.

The nest failed. When there is no male present, the instinct to breed can be so strong that two females will pair up, which is probably what happened here. This page all Hamurana, near Rotorua. September.

The black swan's nest can be a floating platform built from a variety of plant matter. Both pages Hamurana, near Rotorua. June to August.

While one bird incubates the eggs, the other patrols the nest area. .

Maintenance of the nest is time consuming. A floating platform of soft green plant matter soon breaks down and sinks. To combat this problem, the non-sitting bird adds more plants to the nest. They are pulled up from under the water and thrown in the general direction of the nest. After several such actions they can be put into place. This plant looks like Egeria densa, *an introduced plant that becomes a serious weed when it is uncontrolled. The swans eat a lot of it, but unfortunately also leave small pieces behind that grow into new plants.*

Changeover at the nest. The birds greeted each other face on, then the approaching bird moved behind the sitting bird as shown above. The approaching bird moved closer and the sitting bird quietly moved off and slid into the water. Overall the female spends more time on the nest than the male does – usually staying all night. During the day they may change places every 3-5 hours.

This nest was about 23 metres away across the water. I was too close! The swan glowered and hissed at me but stayed in place.

Newly hatched cygnets taking their first peek at the world. There are more hidden underneath the fluffed out feathers.

What seemed like a good viewing platform.....

..... became a roller coaster a few seconds later.

The cygnets remain on the nest for one to two days before being led onto the water by their parents. The babies shown on this page are therefore newly hatched yet were quite active when not resting under the warmth of the feathers. The one shown on its mother's back was clearly not meant to be there. She gestured to it several times but couldn't quite reach it, so eventually stood up in an apparently deliberate move to tip it off. When she stood, several more cygnets hidden away under her feathers came into view.

RAISING THE FAMILY

The young swans or geese may be cared for by their parents in a small family group, or may gather into a larger extended family where a creche of young are looked after by a pair of adults. This care includes assisting with feeding until the cygnets' necks grow long enough to enable them to feed themselves.

Cygnets imprint on their surroundings, their parents and other swans nearby. This is very important to their well-being and they will die if removed at this stage.

Young black swans fledge at about 3 months and at first have grey-brown feathers before moulting to black as they mature. They begin to fly at about 5 months and can be independent soon after. Some family groups may stay together for some time until the parents are ready to breed again. In this case, the young birds do not start to breed but are tolerated as subordinates and may eventually be chased away from future broods. Their overall life span is about 40 years.

Newly hatched Canada goslings feed themselves but stay close to their parents and also imprint on the members of their family group, which may include up to 50 young in a creche situation.

Immature birds have a grey tinge to feathers, the fringing is less distinct, the cheek patch not as white, and overall the feathers are narrow – only half the width of adult feathers.

Domestic goslings fledge at about 7-8 weeks but stay within the family group for about 10 months or until the next breeding season. They start to breed at about 3 years and, like Canada geese, live for about 25.

Typical travelling configuration, one black swan parent leading the way and the other watching from the rear with the family in between. Gordon Carmichael Reserve, Tauranga. July.

Black swans spend most of their time swimming, but sometimes come ashore for a nibble of grass or a rest. The adult swan largely ignores other bird species unless they come too close. Gordon Carmichael Reserve, Tauranga. June.

The cygnets lack the long neck that allows the parents to reach food well below the surface, so they reach for clumps to give to the babies. This plant looks like Egeria densa. Lake Rotoiti, near Rotorua. June.

Young black swan cygnets are a plain pale fawn-grey colour. The neck is already clearly that of a small swan. Gordon Carmichael Reserve, Tauranga. July.

Canada goose goslings are quite yellow when first hatched. Lake Okareka, Rotorua. November.

Whether on land (above) or water (below), the Canada goose family on the move has one parent in the lead and the other bringing up the rear. These goslings are a little older than those on the previous page and are fast losing their yellow colour as the grey under-down replaces the original yellow.

This group feeding on pasture had one parent standing guard while the rest of the family relaxed and grazed. The birds at this urban lake are so used to people they weren't bothered by my presence.
All this page: Lake Rotoroa, Hamilton. November.

Mixed breed domestic geese show a variety of colour patterns in both the adults and the young goslings. Those above (Katikati) were tame birds that kept an eye on me but weren't overly concerned about my presence. Those below were in a roadside farm paddock at Sun Valley, near Rotorua, and were quite nervous. At first the goslings continued feeding while the adults watched me and started to move away. Then the family was deliberately led away to a safer distance. This page all October.

Canada goose goslings spend time resting on either land or water, always with at least one parent close by. The parents look after their family's safety, but the young generally feed themselves by grazing on land where the food is within reach. These birds are just starting to grow their first few wing feathers. This page: Gordon Carmichael Reserve, Tauranga. December.

Young black swans are more water-based than geese and as their necks grow they are increasingly able to feed themselves. Above: *These birds are about to lose the last of their down in favour of the new feathers showing through.* Gordon Carmichael Reserve, Tauranga. August. Below : *The first full set of feathers mixes a light grey-brown colour with black. These immature birds, stay close to at least one parent whether on the water or resting or grazing on land, even when they reach adult size.* Below left: Hamurana. October. Below right: Lake Rotoma, near Rotorua. November.

Feral geese and ducks at tourist spots soon learn that tour buses contain people with food. In this case, the white gander took his nearly adult family to meet the bus, but the tourists just wanted to take photos.

The immature feral geese looked very adult but had a slightly scruffy appearance compared to their parents. This page: Hamurana, near Rotorua. March.

BATHING AND PREENING

Feathers are very important for birds. They keep the bird warm and dry, allow it to fly, and are also used for communication. It is therefore not surprising that birds spend a lot of time caring for their feathers by bathing and preening.

Bathing is done to clean the feathers. Swans and geese are typical of most waterfowl in their bathing routine. They dip their heads and ruffle their feathers so that the water washes over their bodies. This may be repeated for several minutes, followed by a few strong wing beats.

Preening is the bird's way of grooming itself. It checks through all its thousands of feathers; straightening, placing and oiling them. The oil comes from special glands near the tail and the bird uses its bill to distribute the oil along the length of each feather.

During the moult feathers may be shed during preening. The old feathers are replaced by new ones growing through from underneath. These keep the bird fresh and healthy, and are how the birds change colour as they mature and signal readiness for pairing by adopting their breeding plumage.

Watching birds preen allows us an opportunity to see the birds' feathers better and to appreciate the colour and structure of them. After a preening session the bird will often stretch and provide a good view of its wings.

Preening may be seen at any time the bird is relaxed and if done in our presence, it indicates that it is fairly happy about us being there, which is always nice.

Social birds such as these feral geese will often relax and preen together. Hamurana, near Rotorua. October.

This view of the Canada goose preening shows the layers of its feathers. Gordon Carmichael Reserve, Tauranga. March.

A large black swan bathing actively can make a big splash! Gordon Carmichael Reserve, Tauranga. March.

Even hatchlings need to preen. This black swan cygnet is still on the nest and probably only a few hours old. Hamurana, near Rotorua. September.

LAST VIEWS

A few last photos show some of the variety of the lives of these birds. Whatever they are doing, these birds are a great part the world around them, cohabiting with one another and with us.

These birds appear to be two pair of feral pilgrim geese. The closer pair aren't too sure about my presence. Lake Rotorua. February.

The Canada geese and other birds on this urban lake are as used to dogs as they are to people. They keep a watchful distance from swimming dogs, but don't seem overly concerned. Gordon Carmichael Reserve, Tauranga. June.

The serenity of these two black swans contrasts the action shown opposite. McLaren Falls Park, near Tauranga. September.

The Canada geese and other birds on this urban lake are as used to dogs as they are to people. They keep a watchful distance from swimming dogs, but don't seem overly concerned. Gordon Carmichael Reserve, Tauranga. June.

Black swan in action on the water. McLaren Falls Park, near Tauranga. April.

OTHER SWANS AND GEESE

A number of other swans and geese may be seen occasionally in New Zealand. They are usually domestic birds that can be seen on farms or feral birds that have escaped from domestic stock. The exception is the Cape Barren goose which may also self-introduce from Australia (see below).

- **African goose**. Domestic. This is a large heavy goose originally descended from the Asiatic swan goose (*Anser cygnoides*). Males and females are both grey/brown and have a large knob in the front of the head. Up to 9kg and 1 metre high (male).

- **Cape Barren goose** *(Cereopsis novaehollandiae).* Endemic to Australia. Attempts to introduce it in the 19th Century were unsuccessful. The few birds found in the wild in New Zealand are probably captive escapes, although some may have also self-introduced. Likely fewer than 50 birds. 85cm.

- **Chinese goose**. Domestic. This is an elegant long-necked goose originally descended from the Asiatic swan goose (*Anser cygnoides*). Males and females are both grey/brown and have a large knob in the front of the head. There is also a white strain in which both sexes are white. Up to 5.4kg and 1 metre high (male).

Views of a captive Cape Barren goose. Note the yellow-green cere that covers the black bill; the black feet and pink-red legs; the overall stockiness of the bird. This page Hamilton. August.

- **Mute swan** *(Cygnus olor).* A European swan introduced as an ornamental bird in the 19th Century. There are now only a few small wild populations and individual birds, mainly at Lake Ellesmere in the South Island and Hawkes Bay in the North Island. While not fully protected, hunting is only allowed at the discretion of the Minister of Conservation. Probably not more than about 100 birds. 125-170cm.

The mute swan is unmistakable in New Zealand – it is the only all-white swan and is much larger than the black swan. The female (left above) is smaller than the male. She has less colour in her bill and the black knob at the base of her bill is much smaller than the male's. Captive birds, Hamilton. August.

FURTHER INFORMATION

This is a small selection of the variety of books and other resources available to the person who is interested in learning more about these birds.

Arkins, A. & Doel, L. (2005). *Introducing New Zealand birds.* Auckland: Reed.

Ashton, C. (1999). *Domestic geese*. Wiltshire: Crowood.

Ashton, C. (2010). *Auto-sexing geese now being recognized from Shetland to Australia.* Aviculture-Europe. http://www.aviculture-europe.nl/nummers/10E03A10.pdf.

Ashton, C. & Ashton, M. (2009). *Keeping ducks and geese*. Cincinnati: David and Charles.

Birds New Zealand (Ornithological Society of New Zealand) http://osnz.org.nz/.

Coleman, J. *Review of information relevant to the impact of black swan in Northland.* Lincoln: Landcare Research. http://www.envirolink.govt.nz/PageFiles/250/218-Nlrc28ReviewOfBlackSwan.pdf.

Dos Santos, V. (2011). *Impact of black swan grazing and anthropogenic contaminants on New Zealand seagrass meadows.* Hamilton: Waikato University. http://researchcommons.waikato.ac.nz/handle/10289/5717.

Fish and Game New Zealand Council http://www.fishandgame.org.nz.

Gale, T. (2012). The Canada Goose: half-baked or finally cooked? What will be the future for the 'king of game birds'? *Lincoln Planning Review., 4*(1), 4-9.

Handbook of Australian, New Zealand & Antarctic birds. (Vols 1-7, 1990-2006) . Melbourne: Oxford University Press.

Heather, B. D. & Robertson, H. A. (2005). *The field guide to the birds of New Zealand*. Auckland: Viking.

International Union for Conservation of Nature. *IUCN red list of threatened species.* http://www.iucnredlist.org.

New Zealand birds : nga manu o Aotearoa http://www.nzbirds.com/.

Royal Forest and Bird Protection Society of New Zealand. http://www.forestandbird.org.nz/.

Reader's Digest (1985). *Complete book of New Zealand birds.* Sydney: Reed Methuen.

Robertson, C. J. R., et al. (2007). *Atlas of bird distribution in New Zealand : 1999-2004*. Wellington: Ornithological Society of New Zealand.

Selby, I. (2005) *New Zealand poultry standards*. (2nd ed.). New Plymouth, N.Z. : I. and G. Selby Ltd.

Te Papa, OSNZ & Department of Conservation. *New Zealand birds online* http://www.nzbirdsonline.org.nz/

Wildlife Act 1953 http://www.legislation.govt.nz.

www.ingramcontent.com/pod-product-compliance
Lightning Source LLC
LaVergne TN
LVHW070625061126
851857LV00002B/16
* 9 7 8 0 4 7 3 2 6 8 9 3 0 *